Science Conundrum

The Holy Grail of Science

To

The Cosmic boundary

LAGARI A

Dedicated
To
The world of Knowledge seekers

Shifana

The Inspiration behind me.
This book remains incomplete without mentioning
her.

Contents

1   Preface

This book deals with the undiscovered science behind everything. I mean it, *everything,* ranging from the grand cosmos to its grand mechanism with the dreamed scientific unification of known and *unknown* forces.

The book is divided into three chapters. First one deals with Grand theory of everything with unified equation which explains how everything is executed in and around us. The following chapter's deals with reality, cosmic boundary and universal consciousness.

The second chapter holds a dramatic equation to spot the cosmic boundary and it also talks about false realities around us. The last one brings you to the world of consciousness and the universal memory. Taking you to a journey from the primitive understanding of our brain to the most modern findings about our brain and its systems.

So set your intelligence to get the complete unparalleled scientific understanding of our universe.

2 Abstract

The quest for the theory of everything begins when curiosity origins in human beings making them think and observe and it is from where even science has taken birth. Scientists, Philosophers, Psychologists and even Theologists came up with unified theories from their respective fields to explain everything; and which was a failure. As the quest continues and the quest has turned into a challenge which is proposed by modern scientists. Studied these phenomena's only to find the technique of organization of these matters around us by God or Nature (whatever people calls). If god or a central controlling consciousness exists, there must be common laws which designed the working of this universe. Here disclosing the answer of the greatest question of human kind, i.e. the law behind everything including life structure and opening you a Grand unified theory from all the aspects of life which is acceptable to all in varied fields of study. Indulging in a deep study about the behavior of living beings and compare it with other materials. The interesting feature is that whatever is happening around us is directly or indirectly influencing us. The methods mainly include observations, simple experiments and study of various research papers in this field of study. The research is carried out as the balance of the search of many renowned scientists like Newton and Einstein and overcoming the proposals of Gödel and few others. It is finally understood that everything around us is connected to one another by the means of energy waves. It also satisfies the chaos and also tells why some others told that we cannot find the ultimate theory. The work also points on the certain properties of energy waves to maintain everything as it is. The singularity of all that we know and not known yet. Exploring through the channels of science, psychology and philosophy for the complete scientific understanding about the world that we are in.

3   Towards Unification

In early times, during the first discovery of fire, man used his observational skills for understanding the cause of forest fire. He found that the fire is due to the friction which is created by the rubbing of trees against each other during storm. After a long time of observation he used the same rubbing of trees with rocks and there the first artificial fire was born. That fire of curiosity lights up the minds to capture the result that we see and experience today.

On the way of development, man has developed theories, equations, derivations and finally laws. The theories got its birth by observation of man towards him and his surroundings. In the early times, even still now these observed "common matters" are summoned into a statement. Some call it philosophy while some other call it a theory. In the present scenario, philosophy and imaginations are changing to theories and laws. Major case is with the imaginations of recognized scientists as they state those with logical reasoning or by means of mathematical steps or equations. People don't have time to rethink or question as those are said by the educated elite. So it simply sense that when an imagination is put forward by a recognized scientist (in the sense that a man who deals with science) then it is termed as a theory or a truth.

From ancient time onwards man begins to think "what is the cause of everything around him?" he is also sure that the cause is not him as he knows that even before his birth and after his death the world will exist as it is. The question further grows to the first quest for the "unified laws". They finally found answer in externalities such as flower, sun, moon etc.

According to history, God evolved when man started worshipping external objects by giving human features. When man felt sick, he found solution in Neem plant. Then he worshipped the Neem and this method expanded to other objects and finally reached in the creation of a number of Gods. People used to teach the generations about their Gods and worships. This teaching is carried over by stories. The imaginative stories gave God a human nature and feature. Thus the curiosity for the unification of everything lead to the singularity of God.

When atheists arise, they began to question the existence of god and they claimed everything is from nothing as the universe is electrically neutral. All matter that we see is the positive energy of the universe and the black space is the negative energy. They tried to formulate the theory that makes everything as it is. This quest which started from the origin of mankind travels through the greatest minds like Isaac Newton, Albert Einstein etc. and finally stated that such a unification is not possible.

Some scientists believe that, the four fundamental forces had been combined into one fundamental force, at the beginning of the universe (up to 10^{43} second after the Big Bang). Hence a new theory called String theory emerged. String theory is capable to incorporate with each of the four fundamental forces in the universe. According to string theory, everything in the universe, at its most microscopic level, consists of vibrating strings (or strands) each with a specific pattern of vibration. Due to this varied movement patterns of strings, every particle of its mass and force charge is created or in another way everything have a specific string pattern which is different from others. (That is to say, the electron is a string vibrating one way, while the quark is a string vibrating another way, and so on). As per many scientists such a theory will provide a way to access 'the mind of god'.

4 What the Grand Theory must satisfy?

The Grand theory of everything must satisfy psychology (human emotions), science (mainly physics) and all other field of studies. The major challenge is to unify the human varied emotions and activities which is truly unpredictable with the physics as those are certain under certain conditions. The theory must also satisfy wrong as well as correct experimental results. Till now, scientists all over the world stick to four fundamental forces for the unification. They have don't even thought of psychology. The need to consider psychology is because we are using it as a tool. We have developed mind science which even helped the science elites and we also unveiled the mind functions by neuroscience. The complete unified theory must be capable of explaining everything, even from a bit of movement to the creation and function of the entire systems in its accuracy. All the theories explain the correctness only and those theories exist only by proving the correct matters. But Grand theory of everything should be able to tell all the results including the wrong ones as everything includes both right and wrong, lie and truth and so on. It must also be able to tell why certain matters are unpredictable and incalculable. It must tell why certain matters are in certain way and even when why you are reading this text? And for your surprise, it will also tell whether you will continue reading or not.

5 Just a few words

The methodology to find the Grand theory of everything is entirely different from other methodologies that we prefer today. The important thing is that as the Grand unified theory is the base of everything around us. So we have to observe the basic character and basic functionalities. As the theory is meant to everything in and around us, the only way of finding this is by observation and logical viva. The methods and final conclusions must be acceptable to everyone who thinks with logic and not through eyes.

What do you see from the top of a ten storied building in the center of the town which you are residing? Or simply what you see when you look outside? Obviously there were events taking place. People gathers, splits and don't even mind each other. It is like the twigs which are flowing the fast water current of a large river. All twigs are not from the same kind and are not from the same land. They all meet in the river. Some twigs move away from each other while others collide one another. This is a typical picture of a town. The vast river is the town and its fast current is the busyness of the town and twigs are the people in that town.

Genetic studies have shown that we all belong to the same family which have its roots in the major continents and with the major civilizations existed in the past. But in the real world, we are just strangers. For a busy man he himself is a stranger. Imagine that you are walking through the crowded street near your locality. Suddenly a face flashes beside you. You turned back and that face also turned back to you and stared at you. But interestingly you both don't know each other. Then the quest rises, what makes you both stare each other? Why you are attracted towards that person? And why this is not happening in the case with other people whom you have met? The answer is in the hands of the universal law and engineering's which correlate everything.

6   First look to the Universe.

In order to understand the mechanism of the universe and its grand cosmological engineering framework we have to consider the 3D coordinate system of space. There is no problem in taking more than three coordinates as per some physicists. We are generally taking the accepted 3D system for understanding. The entire coordinate is made up of strings. The *fig.1* gives the representation of the perfectly vacuum space. The *figure.1* possess strings which resembles a straight line. That is, the strings are at mean positions. The straight strings shows the vacuumness of space. But this is not real as there is no perfect absolute vacuum existed. We will discuss about vacuum later.

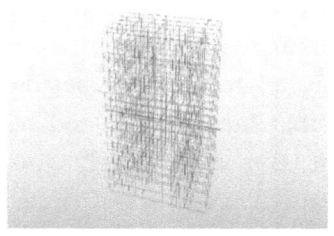 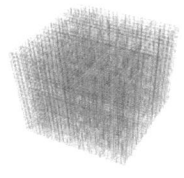

Figure. 1: Cognitor field

 *Figure. 2 (Scan barcode for a simplified animated view of cognitor field)*

The strings are interconnectors between anything and everything in space. It shows deflections according to each matter which is present in it. These deflections and the waves originate due to this deflection from the mean position tells what's present there. These waves which represent the matter (the very existence) is called ***Cognitor waves***. The cognitor waves are different for each and everything in the universe ranging from the smallest particles to the biggest stars. The cognitor wave for absolute vacuum is straight lines (which is not existed) and that of classical vacuum is not straight lines. They possess variations from the mean position. By studying and formulating the cognitor waves in detail, we can clearly and specifically distinguish everything by just comparing the waves. The words in this book constitute cognitor waves and each letters too. And the book is a collection of a large number of cognitor waves. Thus the collection of cognitor waves is called *Fluctus Samaharam*. The entire universe consists of infinitely infinite collection of Fluctus Samaharam and is called *Ekathwam*. Ekathwam is the final cognitor wave of the entire universe. Now you have the blue print and have stepped into the first step towards the unified theory.

The tiniest particle in the whole creation may constitute a bunch of cognitor waves which gives us information about what that is and what its properties are. So to that extent an atom may have lakhs of interconnected cognitor waves or even more or less. These waves will add and subtract each other and give rise to a new cognitor wave which is the resultant of all the cognitor wave operations. The cognitor wave is the basic wave. They are the father of all the forces in nature namely gravitational, weak and strong nuclear and electromagnetic forces. They act directly or tends to act these forces. They possess all these force faces in its need.

The foundation of the theory lies in the hands of Isaac Newton who formulated force. Whether we are discussing to prove and unify the four fundamental forces or we are in need of proving everything and just not only the forces as per scientific tradition, we need to consider the greatest mind who have given force its identity.

Force is carried out when a mass is displaced with an acceleration and it is simply by classical physics, it is the product of mass and acceleration. It is also termed as the strength or energy as an attribute of physical action or movement. The physical action or movement that took place here is change. Change from the state. This change which has given birth to force is an event caused due to cognitor waves. Simply, the change in cognitor waves create force. So force is simply an application or result of change in cognitor wave. Then if we question what makes the cognitor wave to deflect from its mean position to create a force, then we repeatedly ask and ask and finally reach the big bang.

7   Big bang-Cognitor Wave.

The big bang itself is the result of a cognitor wave function. We are not discussing about what cause the big bang as it's from the big bang itself everything including the curiosity to question rose. Just before the big bang, the whole thing is a tiny stuff of energy. Then it expands unfolding itself into the vastness that we see today. Right after the big bang the whole universe is just a stuff of infinitely spread gas.

In 1982 a group of scientists including Stephan Hawking proposed an idea that can call as "The luck of imperfection". According to that, the infinitely spread gas is uniformly attracted by gravity from all sides. Due to some imperfection which occurred, creates a change in the equilibrium of the gas caused by gravity. This defy of gravity caused by the "luck" created the first structures of the universe.

It is quite difficult to believe in the imperfection idea put forward by these cosmologists because to me what we call imperfection is actually perfect. It is our inability to understand the perfectness which makes us to think it is imperfect. Our thoughts are imperfect and not the universe. Anyway if this imperfection is the reason, then there must be a cognitor wave which act for that imperfection and they should answer what cause the cognitor wave to create such imperfection.

We can consider another dimensional view of creation which is far different from those historic concepts. Consider the whole universe in its beginning as a small ball bearing. Select a stage to perform this act of creation. This ball bearing representing energy has a specialty that if it bounces then it will get doubled to matter and antimatter. Each bounce is an inflation. The total mass of the ball get halved too. [See *figure. 3*]

The two balls thus produced get bounced (inflated or blasted) a million times and the whole mass which is created by the energy as matter and antimatter gets distributed to its vast copies. As time passes, the change from singularity to entropy increases. The first initial ball is the big bang and the final stuff of distributed matter-antimatter and energy is the universe that we see now. If the re-bounced balls gets collide, then there forms a new mass which is an addition of the colliding masses or an energy which is the result of matter-antimatter collision. The whole event took place in the stage of cognitor waves. The fluctuations in the wave's speedup the collision rate. As a result, in a small time the whole universe gets spread over in the space. The second law of thermodynamics plays a major role in this creation process.

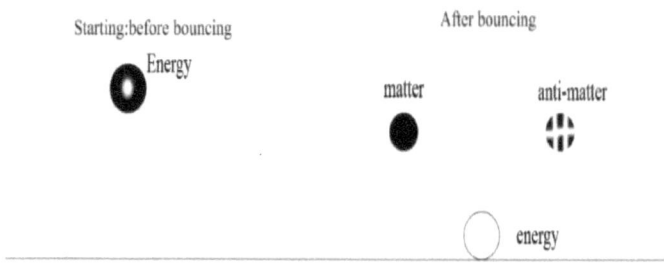

Figure. 3 Illustration of the creation with inflation.

For understanding the vastness, we can imagine a small universe which is so small and having the singularity energy of 10000 kJ involving a 100 events (inflations). Then the universe which popped out from a space with an energy of 1×10^7 Joules. The energy got distributed to 200 quanta, having 50000 Joules per quanta. When these energy get converted into mass then it is valued to 5.555555×10^{-13} kilograms. Here is a picture (*fig. 4*) of computer generated output with these initial values of energy.

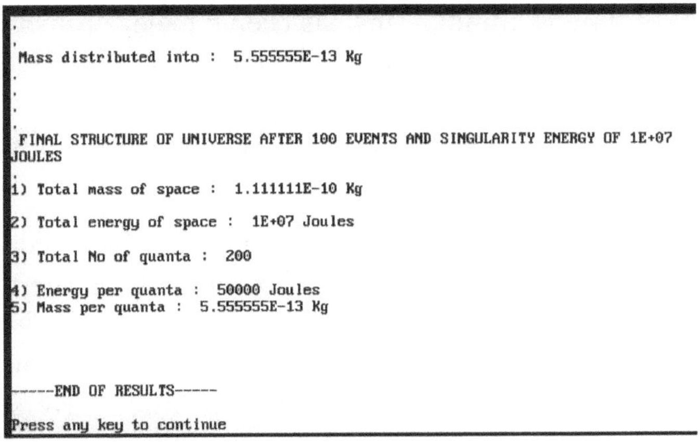

Figure. 4 Computer generated result

For getting even more complicated we can apply the "game of life" devised by the British mathematician John Horton Conway in 1970.

The cognitor wave has its own ways of interaction. These specific ways of wave interaction is the supreme controller in the functionality of the universe. By keeping this in mind now we are ready to understand the grand engineering of the cognitor wave in the creation.

8 Engineering of Cosmic Masterpiece

The origin of universe set the mainstream debate from the very beginning of knowledge itself. It can be found that in the medieval philosophy, the prominent question is whether the universe is finite or infinite. Aristotle stated that the space is finite and time is infinite. This has been a major problem with Islamic, Jewish and Christian philosophers of the middle age who found that the Aristotelian concept of eternal does not integrate with the Abrahamic view of creation.

One of the most prevailing and widely accepted theories about the origin of universe is the theory proposed by a Belgian priest, Georges Lemaitre. He is not only a priest but also serves as astronomer and professor of physics at the Catholic University of Leuven. His theory of the expansion of universe is widely misattributed to Edwin Hubble. He was also the first to derive Hubble's law and Hubble's constant. He published these two years before Hubble's paper in 1927. His "hypothesis of prevail atom" or 'The Cosmic Egg' which is now widely known as the "big bang theory of the origin of universe".

Discoveries in physics and cosmology grew enormously and strive the need of a beginning to the universe. Big bang theory is an effort to explain what happened at the very beginning of our universe.

According to big bang everything pop up in a glimpse of time, everything just begins from nothing at all. At its simplest, it let us know about the beginning from a singularity then inflating over 13.8 billion years till now. English theologian Robert Grosseteste described the birth of universe is an explosion and the crystallization of matter to form stars and planets around us. Many consider this as an explosion but actually it is an expansion.

In 1610, German mathematician Johannes Kepler argued for infinite universe based on dark sky but later Isaac Newton described the large-scale motion throughout the universe after seventy-seven years.

During 1910s Vesto Slipher used spectroscopy to investigate the rotation periods of planets and composition of planetary atmospheres. He was the first to observe radial velocities of galaxies. While Carl Wirtz observed a systematic redshift of galaxies. These observations later account for the fact that universe is expanding. In the same decade Einstein's theory of general relativity based on his field equations said that the universe was described by a metric tensor that was either expanding or shrinking. According to the standard model, the universe has just came into existence from singularity billions of years ago.

Singularity are zones that thought to exist at the core of black holes. Black holes are areas of infinite gravity and all the mass get crushed, closely packed forming infinite density. These zones of infinite density is called singularity.

Our universe is said to begin from a singularity which is incredibly hot and small. Then it eventually expands and cools and finally forms what we see now, including you. The singularity didn't appear in space; rather space began inside the singularity. Prior to singularity nothing exist even space, time and energy. Simply the whole universe began from nothing.

9 Grand Unified Theory of Everything

Many profound theories came to explain everything. The major attempts and the last one on the line is string theory. Still now it lacks the qualities to distinguish philosophy and pure science. String theory plays in a stage of strings. These strings possess extra dimensions which enables it to stretch and form something like a membrane called Brane. In string theory point particles (ideal particles) of particle physics is replaced by one dimensional objects called strings. They try to explain all elementary particles based on quantum states of these strings. Strings are mainly loops and are invisible subatomic particles like quarks etc. The theory is just confined itself to very tiny particles and recognized as a theory of everything as it attempts to include gravity as a particle to explain gravitational force in quantum level. The theory at its first form was five different independent theories but later in 1990 all these five theories are found to be the result of one and different faces of a single theory called M- theory. The theory soon became a hit due to its 11 dimensions including the three present spatial dimension of space and forth dimension of time. The theory holds the need of hidden extra dimensions of space. In order to be a theory of everything string theory produce lots off dualities of current physical theories. But the theory still changing in its structure when the new theories have come and is just unstable itself.

Psychology and psychiatry is still considered by many as an art, pseudo-science etc. widely. Even though a complete theory should be able to explain that too. Science is knowledge stream gained from observations, logical reasoning and calculations of elites (it is actually the result of curiosity) over centuries. In psychiatry the method and approach varies from person to person organism to organism. By the introduction of cognitor theory all the irregularities, anomalies etc. will found to be unified. In that point of understanding, everything is just termed as science. We now understand what all cognitor theory as a complete theory would satisfy and we also know its level of considering everything is not just in words but in action.

The cognitor theory is self-adjusted itself to all past, present and future theories and conclusions. It holds the right as well as wrong ones. More than a theory of prediction it is a theory of explaining everything. Later we will discuss some of its applications in the coming chapters.

The theory is the balance of works conducted by Newton, Einstein etc. and begins with Newton's equations and then combines Einstein's famous equation $E=mc^2$ which connects energy and mass by constant c^2 (c=speed of light). The mass energy relation found by Einstein was a breakthrough in science. Till then both are considered as two different entities but he showed that it is the two faces of the same coin.

For the theory and of everything, we need to explain the fundamentals: Energy and Force.

9.1   Force

Let us again consider the explanation of Big bang. Scientists claim that the expansion is caused due to pressure difference and some other factors. Then we want to ask that what is pressure? The probable answer is force acting per unit area. Then what is force? By Newton's definition;

"Force: Anything that changes or tends to change the state".

It is a good explanation. But what is that anything and what is the basic anything in the definition. Anything is the change: change in surrounding: concentration change of energy.

Then by substituting the word change in the definition of force by Newton we get: -

"Force: Change is that changes or tends to change the state".

This is the action-reaction chain hence formed;

Action↔ Change in surrounding↔ Change in cognitor waves↔ Change in energy

This is similar to the chaotic condition for the change sequence. Force creates action. Action can be classified into two;

- ➢ Movement base.
- ➢ Non-movement base.

Movement base.

Movement base actions are those actions that can be seen (its effects) with naked eye and can be calculated with concrete techniques identically.

Non-movement base.

Non-movement base are those that can't be seen (as a change which can be measured by any instrument accurately) but can be felt sometimes.

E.g. Love, Attraction, Meditation etc.

9.2 Energy

Everything around us including mass is a form of energy. We also know about different forms of energy in our daily life like light, chemical, electrical, potential, kinetic etc. Then it is possible to have a unified energy or a basic energy that cause the creation of other forms. For that we have to move backwards, backwards even before our creation.

The expansion of big bang is considered as the father of everything that we see including us. The whole universe is getting created at the time of Big bang and even the time itself is created by Big bang. Here we solve the unanswered question by the expansion.

During an expanding explosion (consider Big bang), energy is get emitted and after the emission it soon changes into different forms. These changed forms of energy are the ones that we experience today around us. The energy that arise at the time of expanding explosion i.e. at the time of origin of universe changes in forms to accommodate the needs of time.

Here a question arises, then why we have not known about such energy? The answer is simple that is our understanding is limited to the time of birth of "Time" itself. The time have its origin when the light origins. Therefore it is clear that the 'basic energy' called 'Ekaoorjam' is converted or transformed to other energy forms and this transformed energy is resulted in the formation of heat during the Big bang expansion. These energy which formed with heat (unknown ones) are finest of its kind as heat is one of the low quality energy form as per thermodynamics.

We now know where the first energy formed, but what is this "ENERGY"?

The probable answers include the opinion that energy has various definitions. The main two definitions are as follows:

- ➢ Something that have the ability to cause change.
- ➢ The strength and vitality required for sustained physical and mental activity.

As energy is something that have the ability to cause change and then why don't we have the unified energy? It is because from the first explanation / definition of energy, it is clear that energy is something that cause change. After causing the change, that change leads to cause another one. Finally all these chain of changes results in the formation of numerous forms of energy that we experience today.

Relation between force and energy.

"Force: Change is that changes or tends to change the state. -1

Energy: Something that have the ability to cause change." -2

From 1 and 2 we get;

Energy is the cause of force and both appear in pairs.

Therefore;

The functionality of everything around us is summoned to a single statement that:

Each and everything is a function and functional unit of energy.

This is the law of everything. The theory and law of everything is just this simple statement.

9.3   Work

Work is said to be done when some force displaces a body. In physics work is said to be zero when either force or displacement is zero or force and displacement are perpendicular to each other. Simply in physics if you carry some load in your head and walk to your home is not considered as a work nevertheless of your effort. This is because force acting due to gravity is downwards whereas your displacement of load is parallel to ground and is perpendicular to force of gravity.

$$W = FS\ Sinx$$

Force is executed by movement of cognitor waves. Simply work is taken place when there is a displacement of cognitor waves (wave interaction) i.e. energy change.

Hence,

Change in state of energy is work.

9.4   Derivatives

These statements further expands to a set of laws which is their functional properties.

They are:

9.4.1   Each and every thing in the universe generates energy.

We are living in an electronic world. So let me take an understanding example from everyday experience. When any electronic device works it emits electromagnetic radiation. This radiant energy is produced corresponding to the activity which is executed by the object. Cyber world make use of these waves to decode data in the device and process taking place in it. This decoding process is widely known as TEMPEST Attacks. A good engineer can decode your personal data from your pc and even from your atm.

Similar to the electromagnetic radiation emitted from the electronic devices, there is an energy flow in and out of all the bodies. Parapsychologists and spiritual healers call this as aura. Aura is widely considered divine and is attributed to Divinity and believed to be seen only as a result of clairvoyance. These people also come up with Kirlian photography as a proof for its existence.

The Kirlian photography is a technique which uses to photograph coronal discharge and is discovered by Russian electrical engineer Semyon Kirlian in 1939. More than mainstream scientific research it is widely used as a technique to view, depict and analyze aural field by spiritual healers and other alternative medicine practitioners.

Whatever they name it and call. It is truth that there is energy radiation from and towards everybody in the universe. By studying these cognitor waves we can determine the state, nature, condition, ingredients and lot more about a body.

The cognitor wave interaction between bodies create a force. Thus formed force between bodies is called gravitational force (force of attraction) proposed by Newton in his work.

9.4.2   The greater energy (impulsive force) will win (due to greater momentum).

He who has the ability would be the leader (Malayalam proverb)

9.4.3   There is only one thing called energy (for our convenience we termed it as positive negative and neutral)

9.4.4   There is an emitter and receiver between the energies.

The energy radiated from a body get absorbed by nearby bodies or get attenuated in space.

9.4.5   The receiving energy from a body depends upon the nature of the receiver.

Wave mechanics takes place. Sometimes waves get added or subtracted each other resulting in new wave or nothing. The radiating field around the body decides whether an outer wave can interfere the body.

9.4.6   Frequency of energy varies according to condition and nature of body.

The frequency of radiation is influenced by the state of the body.

9.4.7   The flow back of energy varies according to the condition (Health) of emitter, receiver or even both.

9.5   Health of the system.

The state of body depends on certain factors that we can call as Health of the body or system.

Health includes the following;

9.5.1   State of matter.

State of matter includes all known states and unknown states of matter. A solid is defined as a form of matter which possess rigidity. The constituent particles in the solid oscillate in their mean positions. A crystal is made up of atoms arranged in specific order. The order of arrangement tells what crystal it is. A regular 3-D arrangement of particles in space is called crystal lattice. The smallest portion of the lattice is termed as a unit cell (atom). The arrangement is different in liquid, gas and so on.

9.5.2   Color of the body.

It is a result of electromagnetic waves and how brain perceive it. The decoding range of waves of color varies from organism to organism.

9.5.3   P-factors

These are factors that is applicable to living bodies that have ability to think and decide, example: humans. This is sub divided into

9.5.4   Psychology

9.5.5   Data analyzation of brain

9.5.6   Response to both mental and physical stimuli.

9.5.7   Motion of the system

If we are considering a mechanical system (any system can be considered). Then the internal motion executing by the system also influences the cognitor wave radiated as part the system. In case of human body the motion of ions, blood, vitamins, food, air etc. are accounted.

9.5.8   X-factors

These are all other factors which affect the radiation of energy from and into a system.

10   AURA, ENERGY AND SCIENCE

You may hear of people healing others through their looks and certain hand positions (Reiki). Sometimes they predict your future, past and all about you. How these 'black magic' possible? Is it true? Science is here to reveal it for you.

Reiki and other healing and predicting techniques are based on energy, namely Aura (bio energy). Aura is a seven level bio energy field that gets activated when the body stabilizes the energy inside itself by concentrating. Science proves that the body which is not concentrated is actually distributing its bio energy to the surroundings. The bio energy is the energy that is stabilized in our body. The concentration help to stabilize it to a point i.e. to our body itself and hence the wasting of energy to the outside is not happening and hence the energy rise as it is not distributing to all parts. The first stage of concentration (meditation) involves stabilization of energy for the formation of Alpha state followed by Beta and Gamma in the following stages.

The capturing of these waves which got emitted from the body help us to study about that the behavior and every aspects of that particular body. These normal waves move like a light beam whereas a wave from the concentrated body moves like a laser beam. Therefore concentration can act as a better force to create a notable change.

11 NEUROLOGY AND PSYCHOLOGY

As per a recent research studies each thought is associated with each breath. The concentration involves concentrating our thoughts to one and later in a state of zero thought. (Zero thought doesn't mean that the body is not having breath. Breathing is an easier way to get merge into single thought.) It can be done by concentrating in our breath. This is clinically proved and practiced by millions around the world.

As we concentrate in one thought and later merge into a state of nothingness. As per the neuroscientists thoughts are patterns of electrical activity inside our brain. This electrical activity can be mapped into a frequency graph and we can identify the cognitor wave of that particular thought. This is where science plays. The feeling of nothingness is due to the role played by the neurons in the brain. The brain opens its neural channels. In the case of a normal being neural channels are active for only a certain areas, the areas where he is concentrating more in his lifetime. I.e. the active neural channels of a math teacher is entirely different from that of the active neural channels possessed by a musician. But the concentration of energy to oneness involves opening all the neural channels and making the whole channels in our brain active. By concentrating we are allowing the energy to activate the neurons even though we are not utilizing. This increases the efficiency of the brain. Therefore it is suggested to meditate for many health problems and even the 'cool off' time before exams.

12 MIRROR NEURONS AND REALITY

We all see movies, hear music and have other entertainments. Do you know why we are entertained when we hear a favorite music or attend an event? The entertainment itself is a science. It is not depends on how the movie projector or the I-pad works. Nothing becomes an entertainment until we enjoy it. Then how are we enjoying? The enjoyment is a feeling that we are also the part of what we are indulging. Seems to be philosophical right? But science is behind this philosophy. Our brain possess certain neurons called "Mirror Neurons". These are the channels which help us to access others mind and enjoy what others have. Consider;

A man is playing game in his computer. We can see that the man is behaving as if he is inside the game and the game seems to be calling the man to play it more. We feel he is also a part of the game. How it is possible? Both game and person are entirely different ones. But he is in the game. This is where the mirror neurons play their role. It is brain which receives and processes the images and sounds from every part of the body and the brain is network of millions of neural connections. The brain try to create the same neural connections that is responsible for the enjoyment and this is done by the neurons by creating the same connections that the creators intend to connect and to feel. If the man's brain is not working properly the ability to enjoy or experience is not fully get into action.

One of the reasons is neuroplasticity. Recent neurological research has confirmed the existence of empathetic mirror neurons. When we experience an emotion or perform an action specific neurons fire, but when we observe someone else performing this action or when we imagine it, may of the same neuron fire again as if we were performing the action ourselves. These empathy neurons connect us to other people, allowing us to feel what others feel. And since these neurons respond to our imagination, we can experience emotional feedback from them as it came from someone else. The system allows us to self-reflect. Here is an experience of one of my study material.

"Sometimes I used to have some romantic dreams. Among one I saw me kissing a girl. The interesting fact is that I suddenly wake up as kissing is just unacceptable for me. But even after when I wake up I feel the taste of her lips in my tongue. When I lick my lips I can taste her lips":-. How is that possible? Even after a month he can feel it whenever he licks his lips.

It's this mirror neuron which makes us cry after some movies and the same ones make us proud and patriot with some other movies. We have developed technology that enables us to read human mind. The computer is programmed to function as our mind says. The computer here analyses brain waves to decide what the need is and what action to be performed. These brain waves also constitute its cognitor wave. Like mirror neurons we can recreate the incidents of ones memory to others by simply recreating cognitor waves.

13 RESULTANT

13.1 Everything is made of energy.

13.2 The basic energy caused the basic force.

$$\because \frac{dE}{dx} = F$$

I.e. Force is change in energy with unit displacement.

13.3 Everything is connected to everything else.

13.4 The connection is carried out by cognitor waves.

13.5 Cognitor waves varies each other and act as a cosmic identity for each body or system.

13.6 Even the tiniest particle or disturbance contains more or less several cognitor waves.

13.7 Creation and its expansion is a result of cognitor wave action.

13.8 The basic energy and basic force constitute the all forms of energy and force that we see today.

13.9 The workings of energy waves are responsible for the events that are created around us.

13.10 It controls the physical and psychological world.

13.11 It proves everything despite of good, bad, wrong etc.

13.12 It satisfies every set of answers and future discoveries.

13.13 Cognitor waves of living body can be used of studies and varies test and even in curing diseases.

13.14 The grand unified equation.

$$E_f = \sum_{i=1}^{n} f(e)$$

Or

$$E_f = \sum_{i=1}^{n} e$$

Where E_f is the final event, f (e) is the factors or function of that event that affects the event E_f,
The final event E_f or final result is the sum of all the events or factors that affect the event.

14   APPLICATIONS OF COGNITOR FIELD THEORY OF EVERYTHING.

APPLICATION: 1

14.1.1 REALITINESS OF REALITY: existence of universal boundary

We term anything and mostly everything that we experience as reality. Reality is a state of believe that the 'Matter (reality) 'had taken place rightly. It may be an incident or anything ["matter" varies]. We are entertained a lot by the reality shows. Actually are they are all real? We only know about the reality that is shown in the 2D space of TV. Each scene is filmed by a specific script. And we only know the script and we believe the script as reality. We are unknown about the matters outside the script and outside the T.V screen. So we believe that there is nothing outside 'reality'. We calculate reality as those we see and feel by our naked eyes. Then what makes us to see the reality. It is light which enables us to feel and see reality.

According to Indian ancient scripts, Vedas and Naadi Shastra, reality is the time of existence .They counted the existence as a time called Naadi. One Naadi is represented by many ways as most people don't have a real picture about Naadi. Naadi is the time between one complete exhale and before next inhale. Therefore, the reality exists only for a period where the air we breathe (pranavaayu) not is inside us. If we can believe them, then there exist about 14 realities in one minute.

Reality is a word with present tense. It is the very moment of existence. As per physics also the present tense only have a short life span. And past have a lifespan which is increasing. Future is only a curiosity about the next present and there is no future tense actually exists. We perceive reality with our well-defined senses. These senses enables us to be in the sphere of reality. The major form of existence is considered by vision as our motto is seeing is believing. In order to satisfy the purpose of seeing we have eyes. Eyes won't alone make us to see. For seeing we need the major ingredient light. Before going we must know what is light and how reality becomes the phenomenon of light.

We can represent the past, present and future by a light cone. As every event in present takes place or seems to take place because of our feelings. This feeling of experience is created by the light and which is sensed by the brain to term as reality.

The *figure.5* represents the Einstein-Murkowski model of space time. In the *figure.5*, the point P represents past and F represents the future. This idea is better understood and also explained by the ripples in water.

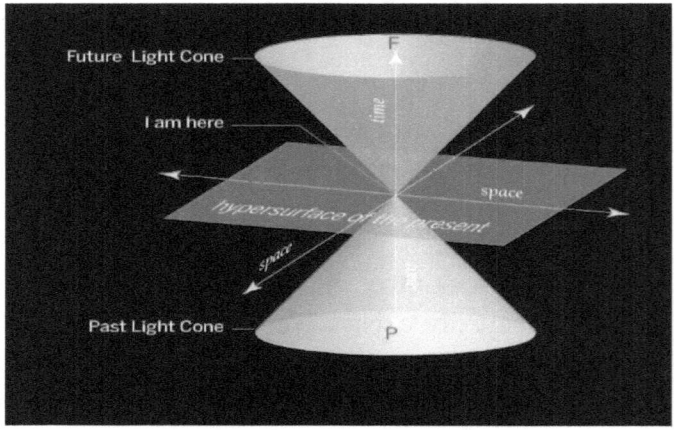

Figure. 5 Einstein-Murkowski model of space-time

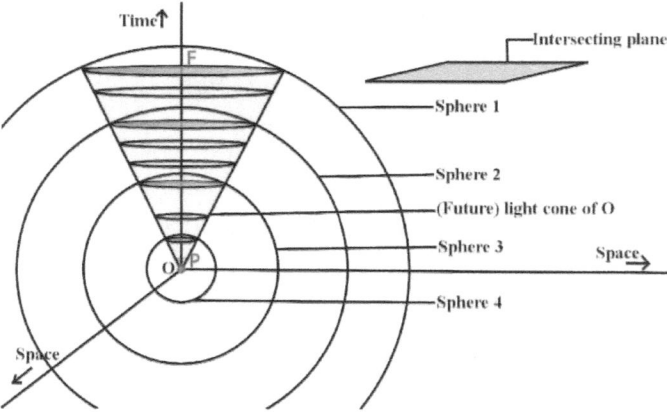

Figure. 6 Light cone represented by ripples in water.

In the above *figure.6* the P[present] represents the time when a stone hits the water body. The ripples firm and create a cone. The point F represents the future.

The point Q in the *figure. 5* shows the area which is not present inside the light cone and thereby not exists. This is because the points outside the light cone are not calculated as light is not going through that part. Here is a case. Consider that you are searching for your pet in darkness inside your room with the help of a torch. The light from the torch moves same as like the light cone. In the darkness of the room, you are pointing the torch to many points. You can only see the things inside the light cone of reality and according to that, those things that are inside the light cone of your torch only exists and the rest didn't. That means as the torch is in your hand and the light is not focusing at you.

Therefore you also didn't exist in the frame of reality.

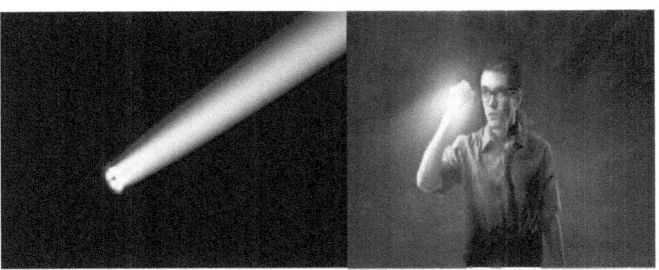

Figure. 7 Area of reality

The only way to avoid the inexistence due to the in appearance in the field of light cone is that to spread the light to every side in a way that it falls on all bodies which is need to be in existence, or in the frame of reality.

Let's do that

Figure. 8 Merging of two realities.

In the above *figure.8*, the light fall on everything and it falls back to the source itself by its property of absorption and re-emission, to make the source also in the frame of reality. The points R and S show the intermixing of individual light cones.

Now it's time to rewind the *figure. 6* which shows the ripples model of light cone. Ripples are the waves created in the water. Like that, light is an electromagnetic ray which possess both particle and wave nature. We are here taking the wave nature of light as ripples in water. Is water simply consisting of a single wave?

For understanding the idea clearly we have to imagine a pond during rain. Each raindrop makes each light cones in the water [as per *figure. 6*]. The rain doesn't consist a single drop, it consists of a million and millions of drops in which a few thousands will fall on the pond depending on the size of your pond. Consider two drops A and B falls in the nearest points in the pond [*figure. 9*]. The light cones created by each drops gets mixed and the wave get added and cancelled.

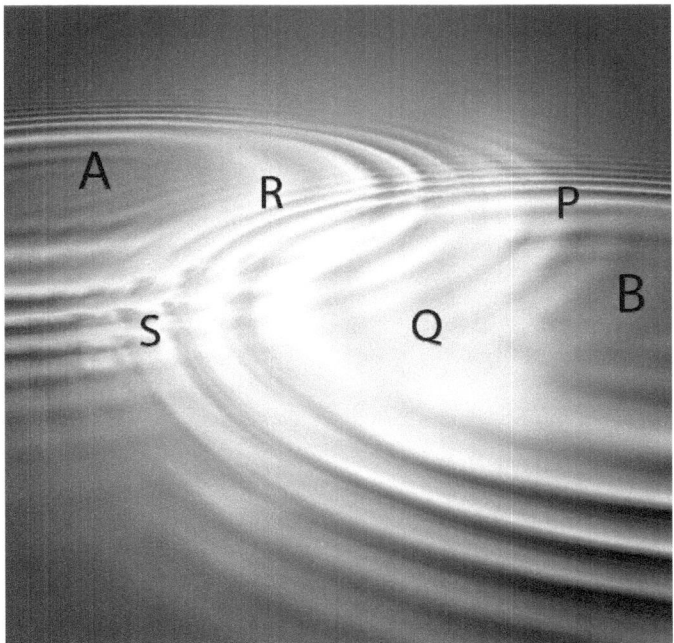

Figure. 9 Two drops hits in a water body

The points P, Q, R, S and V are the points of coagulation of waves which is created by the drops

A&B and called false points. These false points create a coagulation area of PQRS, which represent false reality. Due to the coagulation, the reality which is taken by each drops A&B get cancelled or added and gives a new reality. The new reality may be a sum of reality of A and B or not a reality due to cancellation of waves of A and B.

Now we can go deep into the above phenomena with replacing the water and drops by sun.

W

X

Figure. 10 light cones from sun (illustration)

Sun is a star with spherical and circular in shape. Mathematically a circle is made up of infinite number of points. This infinite number of points creates infinite number of light cones. Thus it is enabling us to see the space around the sun, including us in the earth. But when infinite number of light cones originate, there exists infinite number of false points and false realities. Therefore the reality which we experience is the culmination of false realities. The realitilessness of reality is formed or happened in the case of a distinct star. In fig 10, it illustrates the sun and its few light cones among the millions and millions of light cones. The points W and X represents the area of False realities. As the light cone develops, more and more false point's got created in the universe. Consider a star which is 100 light years far away from earth. If the star dies in 2013, we will only know about that in 2113, as the last light from the star reaches earth by travelling a distance of 9.46×10^{17} in which is 9.46×10^{14} miles! And $2.353702229 \times 10^{11}$ time the perimeter of earth. Science also explains the big bang creation by the means of light cone. The *figure. 11* shows the light cone representation of big bang.

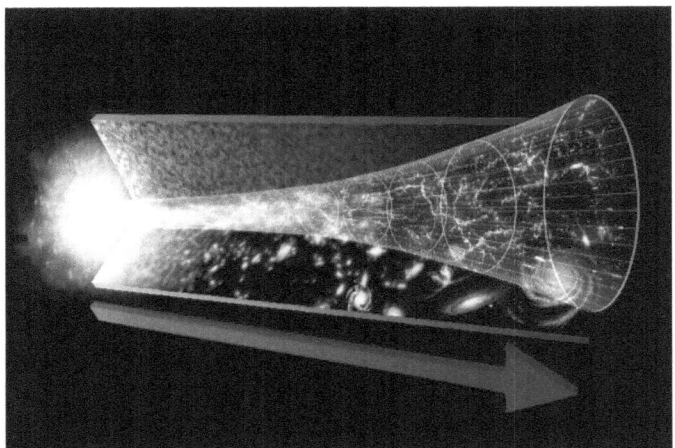

Figure. 11 Light cone representation of Big Bang

According to the fig 1.7, a question arises here is that the light have its birth long after the big bang. And when we use this light cone as a reference of study we will not be able to get the knowledge about the real big bang and calculate its date of birth.

Without intending to question the light cone concept of science as science only exists by means of concepts and theories which can satisfy the general and majority. Science also always put forward a great twist i.e., exception in everything.

14.1.2 Brief-biography of light

Now let's discuss the light cone in the true manner of science and in a different manner of logic. We begin with the birth of light. When sufficient energy is supplied to an atom, the outermost electron of the atom jumps to the exited state. This time, as the potential energy of the electron that jumped to the new orbit increases and which lead instability in that shell. In order to retain the stability, the electron jumps back to the ground state from its exited state. The energy which the electron absorbed when it jumped to the exited state from ground state gets liberated during its journey from exited state to ground state. This liberated energy is in the form of energy packets or quanta. In the case of light, the quantum of energy is called photon. Thus light is produced.

Now we know about the production of light by the jumping of electron. Light consists of a lot of frequencies and only the visible spectrum of light is from 400 to 750 nanometers. This spectrum of electromagnetic radiation was obtained by the hydrogen spectrum experiment. The *figure. 12* shows the various types of electromagnetic radiations which differ from one another in wavelength and frequencies.

From the *figure.12* it is clear that the visible spectrum of light is too short when compared to the whole spectrum of light.

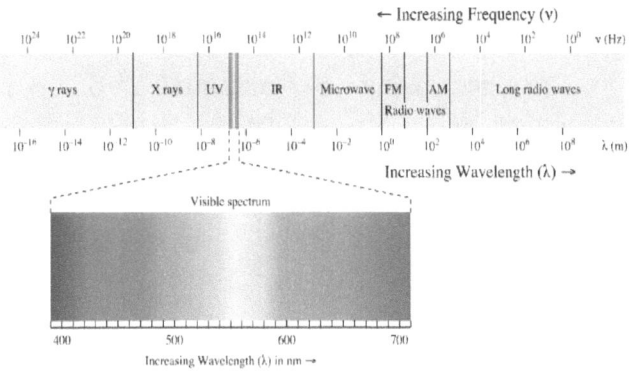

Figure. 12 EM wave spectrum

14.1.3 String waves

All waves possess a velocity, simply wave velocity. It depends on the crusts and troughs it creates. In each oscillation the wavelength and amplitude of the wave decreases. For getting this better, consider a string of guitar. When the player touches the string of guitar a wave is created and thus produces sound. Within a short while, the sound ends and the string comes to rest. At this moment the string have its crusts and troughs in line of origin. This is because of the decreasing amplitude of the string due to oscillation. The *figure. 13* illustrates the dissolving of waves.

The decreasing of amplitudes of strings is represented in string cone. All strings posses' string cones. In the above string cone *figure. 13*, the point A is formed first and as time goes on point B is formed.

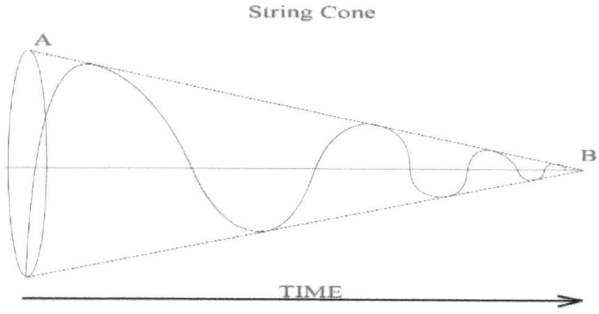

String Cone

TIME

Figure. 13 String cone

String cone is applicable to all waves whereas light cone is applied to path of light from the origin. As light is an electromagnetic wave, and posse's particle and wave nature. Thus the light possess both light cone and string cone. The light cone shows the paths which light sweeps whereas the string cone shows the nature of light waves. The *figure. 14* below represents the light cone and string cone of light with respect to time.

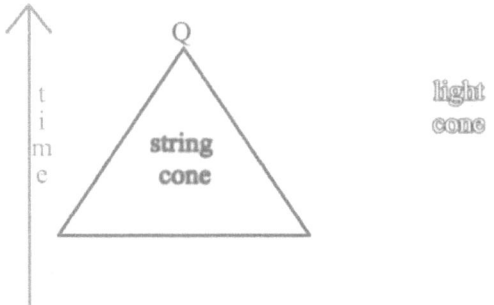

Figure. 14 Light and string cone with respect to time

The point Q in the *figure. 14* shows the start of waves that from that point string doesn't have a wave nature (as per our eyes). Therefore at Q there is no wave or the waves emerge to the surrounding space. At that point the wave loses its ability to continue its journey as a wave.

Now we can insert the string cone in the light cone as per the *figure. 15*.

As per the *figure. 15,* we can understand that when light cone expands the wave nature [amplitude decreases].At point P when the light cone expands to its most, the light waves emerges to the space. Thus the lights have its death at the point P. The decreasing of amplitude of wave of light is to keep the velocity of light a constant.

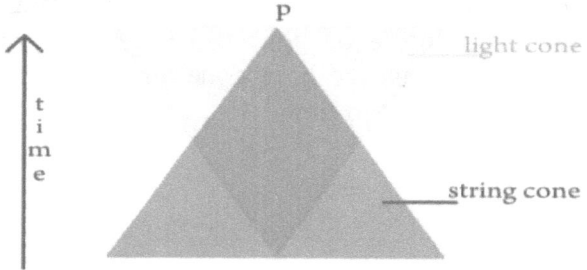

Figure. 15 Superimposing light and string cones

If the amplitude is high and constant, then on moving, the velocity decreases due to its motion. So waves try to decrease their amplitudes to keep their velocity constant [in the case of light wave]. This simply sounds that the light has its death after its long journey.

14.1.4 Universal Boundary

The realities also get dissolved in the pass of time. Now we can take the universe, the land of varied realities into our study. The universe is an endless bordered space with stars and other celestial bodies.

According to string cone, the wave of light only can travels to a specific distance. If the universes have a boundary outside that distance, then we people can't able to know about it and thus we think that the universe is borderless. And the light from the border of the universe will not reach us. We then see dark everywhere. See the *figure.16*, it illustrates the universe and its boundary. The circle A is the boundary of observable universe. The circle B is the boundary of real universe. The circle D is the boundary line at which the light from circle B dissolves in space. The space C which is between the circles A&D is the space that our universe exists in. our universe should expand a lot to see the boundary of space.

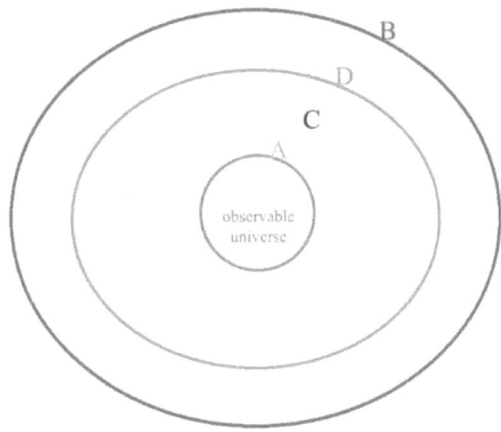

Figure. 16

The light cone illustrates the path of light and string cone illustrates the wave nature of light. From that extend by analyzing the *figure. 15* we can say that as time increases amplitude decreases and finally went to zero. Zero condition occurs at point P and at that point the light have swept completes maximum area.

By applying the boundary condition in *figure. 15* a new equation is formulated to calculate the space boundary. The equation provides the distance a light ray travels before it dies out. When considering the universe as whole the total intensity is to be calculated and that intensity determines the universal boundary.

$$D = \sqrt[4]{\frac{3C^3K}{\pi\lambda^3}}$$

'D' is the total distance travelled by light before it completes it life journey. 'C' is the speed of light. 'K' is a constant. 'λ' is the wavelength of light.

Figure. 17 Scan the barcode for complete equation derivation

The distance 'D' is called boundary distance and if we observe beyond this distance we can't able to see or experience the reality which ends in the boundary. If we calculate the answer will go more than one septillion. (a digit with 24 zeros). The boundary of a single red light is at a distance of One trillion three hundred thirty-six billion nine hundred seventy-three million nine hundred six thousand fifty-four

14.1.5 Resultant

It is found that the reality that we claim as reality is not reality. It is a mixture of false realities. We all don't exist rightly as everything is false reality and mere imaginations. This also questions our existence in the sphere of reality and questions the difference between reality and dream as most times we got confused with realities and dreams in our daily life. As we don't exist, life also not exists, that means after life also didn't exists truly by science.

It is finally concluded as follows;

- ✓ There is nothing called reality exists.
- ✓ Everything maybe a mere dreamlike matters created by brain and our consciousness.
- ✓ There is a visual boundary for the universe.
- ✓ The light also die (attenuates) after some time.

APPLICATION: 2

14.2 DNA CLOUD COMPUTING:
World of universal collective
consciousness.

Every human being has a "key code" in their DNA which helps or influences the behavior of the bio energetic waves which is emitted from the body of that living being. All living organisms (those have mental system and the ability to recollect) are storing their experiences and mental emotions in a cloud server and uses cloud computing technology in the nature. The key code in the DNA helps to enter into their stored data. We describe some impressive features of biological data storage by the nature, and speculate on approaches to research and development that could benefit the understanding about our memory.

14.2.1 OBJECTIVES OF THE STUDY

✓ To study about the memory storage systems in living beings especially humans.

✓ To study about different systems those are participating in the process of storing memory.

- ✓ To study about the bio energetic waves and its function in storing the memory process.

- ✓ To explain scientifically the Psychic and meditational methods in a more precise manner.

14.2.2 FROM THE PAST

When we dive into the helical structure of DNA through the mega project named Human Genome project (HGP). The project launched in 1990 and last for 13 long years and finally ended in 2003. The project got funded by many countries including US, Germany, China, Japan etc. It is considered as one of the greatest project ever conducted in the field of biology. The project aims to identify approximately 25000 genes in DNA. It also finds the secret behind DNA, 3 billion chemical base pairs. The discovery is widely proclaimed as the cracking of god's code. The project has given birth to new field of study called Bioinformatics.

In earlier days when the people are completely unaware of the god's code. They use similarities and differences to solve paternity issues. As this method is based on similarities and mainly females got the burden in their hands. Many loss their lives due to this unscientific method of finding parents. All the beliefs about birth and sex determination of a child got buried when HGP sequenced genetic codes and it got wide acceptance. New methods emerged for paternity determination such as DNA finger printing. Today DNA hold place in highly protected security devices and checks.

After 1950s many theories prophesized the time travel ability of DNA and they believe that we can recollect the lost memories of our ancient ancestors. Some researched in 1960s claimed that we have wormhole to our past experiences in our DNA.

Studies and researches are taking place to cure diseases from the genetic level and to create a disease free world in the near future. Genetic engineering also promising the dream world of new species and super humans. Many researches are carrying out to find the past live and to prove reincarnation based on genetic memory.

Films got produced taking these themes and last man on the line is movie in Tamil language titled "Eyam Arivu" starring Surya also deals with genetic memory and retrieving of this memory. The film also tells that the genetic memory retrieving help to get back even the memory that the person gained by his lifetime.

14.2.3 BLUEPRINT, MEMORY BANK, INNER SPACE

The DNA within all living things is the blueprint for what each organism becomes, subject to the environmental influences that can also have significant effects.

For humans, recent discoveries about DNA are rapidly changing our views about the importance of this material. DNA may affect us much more significantly than we imagined. And, it may hold keys to further discoveries.

It has long been known that our physical appearance is determined by the combination of DNA from our mother and father. Now, researchers are confirming that certain diseases and disorders have direct links to our DNA. Our health may be programmed to some degree by our genetic history.

Our IQ and aptitudes, musical skills, athletic ability, even our psychological and emotional traits may be significantly affected by the DNA within us.

It has been demonstrated that experiences necessary for survival of a species are learned and that this knowledge is passed on to subsequent generations. In some cases this is most likely at least partially through DNA and the unconscious "instinct" that results. Even tiny and simple organisms learn crucial survival skills and pass these on.

For humans, with our relatively complex brain, feelings and memories, what other kinds of experiences might be saved in our DNA over the many thousands of years when our ancestors were born, lived and died? And, can they be accessed by us here and now?

14.2.4 WHAT WE KNOW AND DON'T KNOW

Scientific researchers are gradually uncovering the secrets of our DNA. They have identified the functions of and relationships between some of this material. Many genes remain a mystery and their purpose is unknown.

Sometimes, these mystery genes are called "junk DNA." According to some researchers, this may be an inaccurate label. Because the purpose and nature of this DNA material is not understood, it certainly does not mean it is useless junk.

As is often the case in scientific discovery, the more we know, the more we realize how little we know. Each question answered can raise many new questions.

For some, our human overconfidence and even arrogance can sometimes trick us into believing that we know all of the answers.

However, in the field of genetics research, there seems to be so much that is not known, that for an open-minded person, these kinds of theories about deep DNA memories cannot be ruled-out.

To conduct our own personal research and to find out for ourselves, maybe all we need to do is listen to our inner DNA.

Listen to the voices, feelings, sights and experiences of our ancestors. Their lives, joys and fears are within us. In that way, they are with us always.

> ➢ How can it possible to store a vast amount of data in a single DNA?
> ➢ Why all researchers are beating around the term DNA?

The answers are that the DNA is not able to store this vast amount of data which include the body system functionalizes, recipe to make a human and also the undiscovered assumptive existence of data about their ancestors also. We knew that even in this ultra-ultra-modern technology period, we do not have a storage device as small as DNA to store everything. The problem not lies in how the DNA can store but it actually lies in how much the DNA can store?

Consider when we went for a movie in a theatre. Even after some many years we don't forgot that movie. We also knew that a movie which is shown in a theatre must have a size above 200 GB depending on its clarity. After storing that movie in our memory we have lost 200 GB of space from our brain or DNA. Similarly as we go through every event that has taken place in our life consumes a memory about trillions and trillions of petabytes.

We know that the small microscopic DNA do not have the ability to store this vast amount of data. We also knew that genetic memory only retrieves our biological basic data and not details about our ancestral and the knowledge and experiences that they gained from their life.

Here we are disclosing the secrets of memory storage and rebuilding the structure without damaging the major foundational findings.

14.2.5 RESULTS

The 'junk DNA 'or the elements in DNA have a special key code. This key code influences all the neural mechanisms of the body and thereby stabilizes the energy which is emitted from the body. This emitted bio energy (aura) is consists of specific frequency which is common to that particular person or group or family.

The whole events that takes place is stored in the central controlling system of the universe in the form of energy which can be accessed. Each individual can access this data with the help of key code which is get transmitted from his or her body through their aura.

14.2.6 CLOUD COMPUTING AND HUMAN DNA

We have a whole bunch of storing techniques or devices today. Recently we have experience a new development of storing data in protein chains. We also have researches that are based on string data in our DNA.

Before going to the vast world of genetic memory, It is prerequisite to get yourself free from the current dimension of viewing and transplant yourself to a new logical viewing.

14.2.6.1 CLOUD TECH

In the field of communication technology we have a new comer i.e. cloud computing. This is same as what is happening inside our world between human beings. In cloud, we are accessing the data which is stored in a particular server from any part of the world. The server contains data about almost everything.

Similarly our memory is not stored inside us. The memories of human beings are stored in a vast storage center of the universe. Like we access the data from the cloud server, we can access the data's from the universal server.

14.2.6.2 THE SYSTEM

A system (here take a PC) is needed for accessing the data's from the cloud. Like that, we are accessing the universal cloud where our data's are stored from a system called human body.

14.2.6.3 THE LINK

We now know about the system which enables us to connect with the universal server called our body. But how this connection takes place? As like a computer, human brain just contain a RAM and a specific operating software called Mind by itself. The system is getting connected to the server by linking of cognitor field.

14.2.6.4 SAFETY SECURE

Now we know about the system as well as the cognitor wave which act as an internet to the server. We know about the system hackers then how could it possible to keep the data's private and not mixed up?

Here is our DNA play its role. The DNA contains a specific password which is used to control and coordinate the wave frequency into uniqueness and thereby the wave from a particular body acts as a password for the particular data stored by him. A good hacker can hack this too.

14.2.6.5 DATA ARRANGEMENT

The data's which are stored in the server is arranged as like a 'cone'. The bottom point 'P' in *figure.18* comprises of the very first folder which have the data of our first memory. I.e. childhood memories. As time passes and age increases more and more folders gets added with specific and nonspecific criteria.

The folders can be divided into two:

➢ Specific: These are those that have specific data's about an event or anything. These folders sometimes have more passwords to get connected by more than one.

➢ Non-specific: These are the folders that have some normal, unspecified data. These are those that are said to be stored in short term memory.

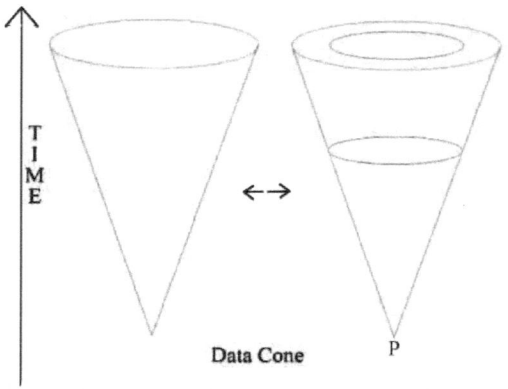

Figure. 18 Data storge cone illustration

14.2.6.6   How this storage is created?

The mega storage system is just the cognitor wave action and its remaining effects. All the events are coded by cognitor waves and these waves get attenuated in pass of time. But there remains signatures of each event. These remains are termed as cognitor fossils. Because of these cognitor fossils we are able to know and feel certain past things when we visit certain places.

14.2.6.7 Mechanism

Each event constitute its own cognitor wave. This cognitor wave act as a cognitor memory bank which anyone can access. We simply copy down our needed details and create our own space of memory. Our brain just have memory like ram of the computer. This memory contains data for the reflex actions and emergency measures. The universal data exists even after ones death and can be accessed by anyone with the key wave. The data is transferred from person to person and person to their or others memory center using cognitor waves.

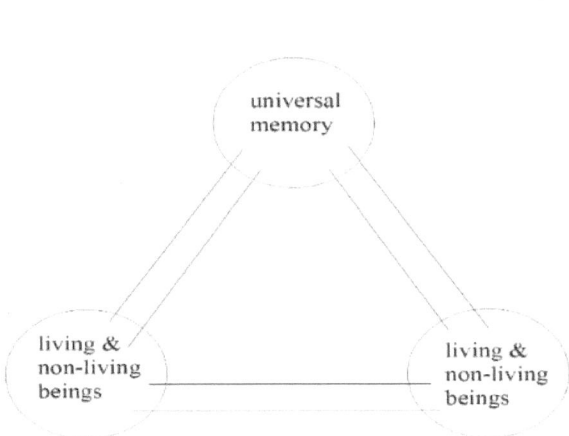

Figure. 19 Interconnection to cosmic memory server from different levels.

14.2.7   Resultants

- ✓ Each aural energy has its own specific frequencies.

- ✓ The total memory is stored in the central server of the universe.

- ✓ The brain only has a functional memory and it is just like RAM in computers.

- ✓ Psychologists and psychic specialists are the hackers (make use of) of others frequency.

- ✓ The energy allows connection between brain to brain and brain to universal server.

APPLICATION: 3

14.3 When Sci-Fi came to reality!!

14.3.1 The dream for Elysium

In 2013 a sci-fi movie released named Elysium. The movie picturizes a world outside earth constructed by humans where there is no diseases and all. It is like heaven. The problem is only high-fi people reside there. All the others reside in earth which is in its extinction. The movie portraits a medical device. The device is capable of curing all diseases.

Now in reality we can construct such a devise by studying cognitor waves and cognitor fields in detail. As each component have its own cognitor wave, if we can able to recreate these waves then the body came to existence from nothing. So it will be a breakthrough in medical field. All diseases can be cured by studying and eliminating the cognitor wave which cases the disease by promoting and treating with an anti-cognitor to nullify the disease wave.

14.3.2 Manipulation technology in the movie Frequencies.

In 2015 a movie released which plays the story of frequencies and its power of manipulation. The in-depth study of cognitor waves help to make the device which is shown in the movie that is introduced as a 'true love finder'. All those manipulations can be down with cognitor waves. In the movie certain words are used to nullify the effect of cognitor waves and to adjust the situation. In reality anything can act as anti-cognitor in order to nullify the cognitor. Words re just one among the many.

15 Consequences.

As we are altering the system of cognitor by creating it in any aspect, we need to face its aftereffect's too. As everything is linked, when any part of system changes it affects the system and also the surrounding. The effect thus created may be small or big. But we have to face the aftereffects.

The further study in the mentioned concept in the chapter "Realitiness of Reality" help us to reanalyze the use of light cones in the varied fields of study and it also help us to explore more about the realities around us. Further exploration in this field give more interesting results.

16 Proofs of cognitor field theory & cloud memory.

16.1.1 100th monkey effect

This is a phenomenon which describes how an idea or data is transferred from a group to another. The theory behind this phenomenon is proposed by Lawrence Blair and Lyall Watson in the late 1970's. The phenomena is claimed to be observed by Japanese scientists. They were observing a Japanese monkey Macaca fuscata for over 30 years in wild.

The scientists served sweet potatoes to the monkey and once in a while a little monkey came to learn itself that washing the potatoes help to get rid of sandy mud in it. The young monkeys teach this technique to heir elder ones. The twist came in the autumn season of 1958 some monkeys among Koshima group observed to wash potatoes. This transfer of idea from Macaca fuscata to Koshima monkey is the observed phenomenon.

16.1.2 Most of have experienced this. When you are in a crowd or maybe with some known or unknown person. Suddenly you can see that he talks or sings what you sing in your mind or what you wish to talk. The brains are connected each other. The data is transferred as thus we externalize our songs through other people. Just like a Bluetooth connection.

16.1.3 The existence of GPS system inside us. The brain by analysis the waves allow us to navigate ourselves.

16.1.4 Near my place there is forest area where the Government plant red oil palm over the area of 3646 hectares. Many domestic animals including goat, buffalo, cow, hen, duck etc. are raised here. None of the animals are tied or care taken by their owner. Animals are freely moving where ever they need to go. Even we humans find difficult to get back home if we're not a resident. All paths looks like and all leads to more deep inside. Interesting fact is that all the animals get back to their owner by evening 6 o clock. When the Asar adhan (Muslim prayer call of evening) is herd all the animals march to their respective place. Farmers don't lose their animal at any circumstances other than theft. These animals are well connected with the cognitor of the forest and we humans from outside not. So we find difficult to get out of the area.

16.1.5 Mind reading techniques (ESP).

16.1.6 When you drive in a busy road and suddenly a horn is heard, you understand what that horn means and we only react to specific horns that are meant to us. Not for the horns that meant for other vehicles near us. Along with the horns our universal memory is helping us to understand the need of those particular horns.

16.1.7 The Global Consciousness Project of Roger Nelson provides experimental proof for the universal memory or collective consciousness.

16.2 So far so now

The world of cognitor waves always work and tries to maintain stability. The proper stableness or ultimate stability is the ultimate vacuumness state. Therefore as we have seen the ultimate vacuum or nothingness is a cognitor field comprising of straight lines. The big bang or the process of creation initiates a cognitor field deflection. In order to attain stability i.e. nothingness, the cognitor field adjust itself and create a counter cognitor wave which we called Gravity. The expanding universe after big bang gets into a uniform level due to the introduction of gravity by the counter cognitor wave action. Similarly corresponding to every action there exist an opposite or counter effective cognitor wave, called counter waves.

Now we can analyze the present day importance of cognitor waves but going through some historic importance and studies.

Many great civilizations have had conquered our earth.to be named the most remarkable ones we have Babylonian, Mayan, Mesopotamian, Egyptian, Indus valley, Incas, Roman and lot more. One of the greatest historians of all time 'Ibn Khaldun' had mentioned about the rise and fall of civilizations in his book 'Muquaddima'.

Every great civilization took its birth from zero knowledge or with some knowledge traces of previous civilizations. Taking the example of Egyptian civilization and the pharaoh era, we can find that extra intelligent brains existed in those days compared to our present day. As per the latest studies it is found that these civilizations had technology and technical knowledge which is more than hundred times than our current technology. This advanced technical brilliance that exhibited by them to the world may have been one of the reasons why people think that pyramids are made by aliens.

When a civilization work hard initially and built a better place they then try to find some place to rest. They become lazy as they have everything ready for them as per the needs. This state is the cause of destruction of all the civilizations. The world need us to be moving continually and effectively. That is our system of existence. Living species urge to new things the curiosity inside them is the key of survival. The advancement attained by civilizations is directly proportional to the depth to which they had to fall. In order to attain stability counter waves destroys the unstable. There always exists a counter action. So as now we are in the peak of development, we should get ready to face the ultimate fall. Maybe our intelligent machines will witness the rise of another civilization after us. The action cognitor wave and counter cognitor wave acts hand in hand to make everything stable. If the action cognitor wave and counter cognitor wave acts simultaneously with time difference perfectly zero, then we can say that nothing had occurred. No event took place. If this condition happens then there would have no creation at all.

16.3 Some extra notes

16.3.1 The cognitor waves can be used to identify, recreate everything.

16.3.2 It act as a universal barcode for everything.

16.3.3 Further study help to create a world without diseases. (The negative impact is there, but we can try to reduce it)

16.3.4 Study help to create life and put an end mark to the phenomenon death.

16.3.5 The characteristic features determine the wave from a body and waves that can accept by that body.

16.3.6 Some researcher's claims an experience in the readouts of electronic devices called random-number generators could be affected by people sitting next to them if those people focused their thoughts on them.

17 Everything in a nutshell

o The Grand theory of everything must satisfy psychology (human emotions), science (mainly physics) and all other field of studies.

o These waves which represent the matter (the very existence) is called Cognitor waves.

o The cognitor waves are different for each and everything in the universe ranging from the smallest particles to the biggest stars.

o The collection of cognitor waves is called Fluctus Samaharam.

o Ekathwam is the final cognitor wave (field) of the entire universe.

o These specific ways of cognitor wave interaction is the supreme controller in the functionality of the universe.

o The cognitor theory is self-adjusted itself to all past, present and future theories and conclusions. It holds the right as well as wrong ones. More than a theory of prediction it is a theory of explaining everything.

o "Force: Change is that changes or tends to change the state".

o Action↔ Change in surrounding↔ Change in cognitor waves↔ Change in energy

o Energy is the cause of force and both appear in pairs.

o Each and everything is a function and functional unit of energy.

o Change in state of energy is work.

o The state of body depends on certain factors that we can call as Health of the body or system.

o Thoughts are patterns of electrical activity inside our brain. This electrical activity can be mapped into a frequency graph and we can identify the cognitor wave of that particular thought

- Everything is made of energy.

- The basic energy caused the basic force.

- Force is change in energy with unit displacement.

- Everything is connected to everything else.

- The connection is carried out by cognitor waves.

- Cognitor waves varies each other and act as a cosmic identity for each body or system.

- Even the tiniest particle or disturbance contains more or less several cognitor waves.

- Creation and its expansion is a result of cognitor wave action.

- The basic energy and basic force constitute the all forms of energy and force that we see, experience, and tabulate today.

o The workings of energy waves are responsible for the events that are get created around us.

o It controls the physical and psychological world.

o It proves everything despite of good, bad, wrong etc.

o It satisfies every set of answers and future discoveries.

o Cognitor waves of living body can be used of studies and varies test and even in curing diseases.

o There is nothing called reality exists.

o There exists a Cosmic Boundary. We can't see or study anything which is outside that boundary.

o There is a visual boundary for the universe.

o The light also die (attenuates) after some time.

o Each aural energy has its own specific frequencies.

o The total memory is stored in the central server of the universe.

o The brain only has a functional memory and it is just like RAM in computers.

o Psychologists and psychic specialists are the hackers (make use of) of others frequency.

o The energy allows connection between brain to brain and brain to universal server.

o Altering of the system creates negative impacts.

o The system is affected by changes in the system as well as surroundings if it is not a completely isolated system with no energy transfer.

o The theory also shake hands with thermodynamics.

o Collective Consciousness (cloud memory) is an attribute of Cognitor field.

o The action cognitor wave and counter cognitor wave acts hand in hand to make everything stable. If the action cognitor wave and counter cognitor wave acts simultaneously with time difference perfectly zero, then we can say that nothing had occurred

18 Acknowledgment

In 2014 January 24th I presented my paper titled "T.A.L.O.E (Theory And Laws Of Everything)", in UGC sponsored National Seminar conducted by the physics department of Iqbal College Peringammala, Thiruvananthapuram. Few months after the seminar I formulated the unified equation. I searched for scientists who are working for the unified laws and found Garett Lisi. He is the one who suggest me to publish my paper as a book. I went on editing and modifying the whole script making it easier for even a common man to grasp. I begin with the basic equations of Newton and further move on to my world of cognitor fields. Now you all have completed your journey along with me in the world of cognitor field.

It's a great privilege to thank my teachers, friends, parents and all who knowingly and unknowingly supported my experiments thereby providing data for my research work. My work has taken long five years for its completion. I am thanking Abhilash, who make me to question the connection between everything. Dr.Biju.S.Padmanabhan (Reiki specialist and psychic researcher) and Dr.V.George Mathew (retired Prof. of psychology Kerala University) for giving me access to people in varied fields of study and sharing their knowledge.

I am extending my thanks to all my physics teachers especially Praveen Bose and Sameer who constantly supply inspiration to do more and my professor Ms.Aiswarya.T (Asst.Prof Applied Science(Physics), Vidya Academy of Science &Technology- Technical Campus) who read and help me with corrections. Also Dr.L.Abdul Khalam (Professor in Physics. Iqbal College Peringamala, Kerala University) who provided me a national platform to present my paper about unified theory. I am also thanking Garrett Lisi (American theoretical physicist) who asked me to publish my paper as a book. Last but not the least I am thanking Subash (English teacher) who work out with the language corrections and my friends Shameem, and Harryz for their help and continuous push to make this book a reality.

19 Reference

No	Title	Author	Year	Source
1	The brief history of time	Stephan Hawking	1988	Book
2	The Theory of Everything (special edition)	Stephan Hawking	1996	Book
3	A lawyer presents the case for the afterlife.	Victor Zammit		Online
4	ToE	Wikipedia		Online
5	The quest for theory of everything	Kitty Ferguson	July 1 1992	Online
6	Dead famous Albert Einstein and his inflatable universe	Dr.Mike Goldsmith	2001	Book
7	What is relativity?	L.D. Landau & G.B. Rumer	2003	Book

19.1 Other references of statements

Item	Source
Definitions and meanings	Google
Other data about history of creation, major experiments, names and works of scientists etc.	From all online sources and documentaries in science.

SCAN THIS TO VIEW THE OFFICIAL BOOK TRAILER